OUT OF THE LAB
EXTREME JOBS IN SCIENCE

ANTARCTIC RESEARCHERS

Emily Mahoney

PowerKiDS press.

New York

Published in 2016 by The Rosen Publishing Group, Inc.
29 East 21st Street, New York, NY 10010

First Edition

Editor: Katie Kawa
Designer: Mickey Harmon

Photo Credits: Cover (sky), pp. 8–9, 26–27 Volodymyr Goinyk/Shutterstock.com; cover (men) Paul Nicklen/Contributor/National Geographic/Getty Images; p. 4 (map) ekler/Shutterstock.com; pp. 4–5 National Science Foundation/Science Source/Getty Images; pp. 6–7 Mark Hannaford/AWL Images/Getty Images; pp. 10–11 Hubertus Kanus/Science Source/Getty Images; p. 13 Dmytro Pylypenko/Shutterstock.com; pp. 14–15 (seal) Yvonne Pijnenburg-Schonewille/Shutterstock.com; p. 15 (penguins) BMJ/Shutterstock.com; p. 16 (humpback whale) Paul S. Wolf/Shutterstock.com; p. 16 (blue whale) https://en.wikipedia.org/wiki/Blue_whale#/media/File:Anim1754_-_Flickr_-_NOAA_Photo_Library.jpg; pp. 16–17 Adam Burton/Robert Harding World Imagery/Getty Images; pp. 18–19 (ice) Auscape/Universal Images Group/Getty Images; pp. 19, 28 (researcher) courtesy of NOAA; p. 21 Ralph Lee Hopkins/National Geographic/Getty Images; pp. 22–23 courtesy of the National Environment Council British Antarctic Survey; pp. 24–25 Vladislav Gurfinkel/Shutterstock.com; pp. 28–29 Konrad Wothe/LOOK-foto/LOOK/Getty Images; p. 30 courtesy of NASA.

Cataloging-in-Publication Data

Mahoney, Emily.
Antarctic researchers / by Emily Mahoney.
p. cm. — (Out of the lab: extreme jobs in science)
Includes index.
ISBN 978-1-5081-4505-9 (pbk.)
ISBN 978-1-5081-4506-6 (6-pack)
ISBN 978-1-5081-4507-3 (library binding)
1. Science — Vocational guidance — Antarctica — Juvenile literature. 2. Research — Vocational guidance — Antarctica — Juvenile literature. 3. Antarctica — Discovery and exploration — Juvenile literature. I. Mahoney, Emily Jankowski. II. Title.
Q149.A6 M34 2016
559.8'9023—d23

Manufactured in the United States of America

CPSIA Compliance Information: Batch #BS16PK: For Further Information contact Rosen Publishing, New York, New York at 1-800-237-9932

Contents

A Cool Career 4

A Frozen Continent 6

The World's Coldest Desert 8

Many Kinds of Research 10

Researching Climate Change 12

Antarctic Wildlife 14

Astronomy in Antarctica 18

Antarctica Rocks! 20

Research Stations 22

At Home on the Base 24

A Day in the Life 26

Becoming an Antarctic Researcher 28

Extreme Science 30

Glossary 31

Index 32

Websites 32

A COOL CAREER

At some point in your life, you've probably been asked, "What do you want to do when you get older?" There are many different careers to choose from. Some of the coolest and most exciting jobs are in science! A scientist doesn't always work in a lab. Scientists work in many extreme places around the world.

Antarctic researchers have one of the most extreme jobs imaginable. These scientists collect data, make observations, and oftentimes live on the very cold continent of Antarctica. Because of the harsh weather conditions in this remote area, this job can be **complicated** and dangerous, but it's also important and can be very fun.

Antarctica

In 1911, Roald Amundsen of Norway became the first person to reach the South Pole, which is located on the Antarctic continent. A month later, Robert F. Scott of Great Britain also reached the South Pole. These two men were pioneers in Antarctic research and exploration.

Many American Antarctic researchers work at the Amundsen-Scott South Pole Station. This station is named for Roald Amundsen and Robert F. Scott.

A FROZEN CONTINENT

Being an Antarctic researcher is a tough job because of Antarctica's climate. It's the coldest place in the world! In fact, average temperatures in Antarctica only reach about −30 degrees Fahrenheit (−34.4 degrees Celsius) in the winter. The temperatures rarely get above freezing, so living conditions are **treacherous**. The temperatures are cold, and sometimes the wind chill makes it feel even colder. Wind chill is the temperature it feels like when the wind blows.

While buildings and labs in Antarctica are heated, researchers often need to go outside to collect data and samples. When they do go out, they must be extremely careful to avoid getting frostbite. This happens when the skin freezes. They must wear special protective gear to stay safe.

SCIENCE IN ACTION

The lowest temperature ever recorded on Earth, which was −136 degrees Fahrenheit (−93.3 degrees Celsius), was recorded in the eastern part of Antarctica in August of 2010. However, this isn't an official measurement because it was recorded from space and not on the ground.

These are some of the layers and pieces of clothing Antarctic researchers need to wear to stay warm and safe as they work in such a cold place.

What do Antarctic researchers wear to stay warm?

eyes
- protective glasses

head
- balaclava—covers the head and neck

body
- many layers, including warm undergarments, or underwear
- a least two layers of pants
- a light jacket
- a heavier, waterproof coat

hands
- at least two pairs of gloves

THE WORLD'S COLDEST DESERT

Antarctica has plenty of snow and ice. At least 98 percent of the continent is covered in it! The ice can be as thick as 15,000 feet (4,572 m). However, Antarctica is technically considered a desert because very little precipitation falls there. In fact, only about 2 to 4 inches (5.1 to 10.2 cm) of snow fall each year in Antarctica. However, the temperatures are so cold that the ice and snow don't melt, so the snow **accumulates** year after year. That's how the ice got to be so thick.

We only know about the cold, dry climate of Antarctica because of the brave men and women who've traveled there to study these harsh conditions. Researchers make it possible for us to know what it's like in Antarctica and other extreme places on Earth.

SCIENCE IN ACTION

Antarctica is the world's largest desert. It's known as a cold desert. The largest hot desert in the world is the Sahara Desert in Africa.

The continent of Antarctica contains about 87 percent of the world's ice. Researchers can learn a lot from studying all that ice.

MANY KINDS OF RESEARCH

If the cold weather and snow haven't **deterred** you from the idea of becoming an Antarctic researcher one day, you're probably wondering exactly what you'll be researching. Antarctic researchers study the climate, wildlife, and geology of this area of the world. These scientists often choose a specific area to focus on.

Since this part of the world is so different from anywhere else, scientists have made many **unique** discoveries in Antarctica. Many countries have scientific research bases in Antarctica. Oftentimes, scientists from different countries work together to solve problems and collect data.

The United States first established a research station, which is called McMurdo Station, in 1955. For over 60 years, Americans have conducted research in Antarctica.

SCIENCE IN ACTION

McMurdo Station is the largest Antarctic research station. It continues to serve as the **logistics** center of the U.S. Antarctic Program.

There are no permanent residents of McMurdo Station or any other place on the Antarctic continent. The conditions are too harsh for people to live there for long periods of time.

RESEARCHING CLIMATE CHANGE

Many Antarctic researchers are interested in the region's climate. Since Antarctica has many areas that haven't been affected by people, it's a great spot to study how weather affects the natural world without human interference.

Researchers are especially interested in looking at the average temperatures in Antarctica to see if they're getting warmer or colder. They've also been looking at the effect melting ice could have on the rest of the world if temperatures were to increase in Antarctica. These researchers have discovered that melting ice from a warming Antarctica could increase **sea levels** all over the world. Antarctic researchers and other scientists want to study when and how this could happen so they can be prepared.

SCIENCE IN ACTION

To study the ice and ground below Antarctica's surface, researchers take core samples. They use a large drill to pull samples of ice from deep below the surface. They can see what the landscape and climate of Antarctica were like hundreds and even thousands of years ago.

The work done by researchers studying Antarctica's climate is very important because changes in this continent's climate could affect the entire world.

ANTARCTIC WILDLIFE

Biologists, or scientists who study living things, have a lot of unique wildlife to study in Antarctica. From whales to penguins, many animal species have adapted to life in the extreme cold.

The Weddell seal is one of the animals most closely studied by Antarctic researchers. Since 1968, these seals have been a species of great interest for scientists. Weddell seals in Antarctica meet to breed, or mate, in the Ross Sea. They get closer to the South Pole than any other mammals on Earth.

The National Science Foundation (NSF) has funded a Weddell seal research project through the U.S. Antarctic Program. Its goal is to discover the impact of environmental changes on animal populations by observing these seals. Researchers involved with this project have studied more than 23,000 seals.

SCIENCE IN ACTION

Penguins are another kind of animal studied by researchers in Antarctica. The Emperor penguin, which lives in Antarctica, is the largest penguin in the world.

Emperor penguins

Antarctic researchers collect data about animals, such as seals and penguins, in order to discover how they can survive in such a cold climate.

Weddell seal

Marine biologists travel to Antarctica to study the whales that live in the water surrounding the continent. Many whales, such as the humpback whale, blue whale, and fin whale, are found in the Antarctic region. Whales eat the **krill** and small fish that live in Antarctic waters.

Researchers sometimes tag whales with special tracking devices that give off a radio signal. They can use that signal to track where the animal travels. Marine biologists use this ability to safely track whales to study their **migration** patterns and mating behaviors. Many whale species are endangered. Researchers from around the world travel thousands of miles to this remote continent to try to figure out how to keep them from becoming extinct.

blue whale

humpback whale

Sometimes a scientist will place a special **microchip** in a whale's **blubber**. This microchip helps them track migration patterns and determine how deep the whale dives. This technology is exciting because it gives scientists a new view of how whales live.

Antarctic researchers often go out on boats in order to get as close to whales as possible.

ASTRONOMY IN ANTARCTICA

The South Pole is one of the best places on Earth to view stars. Because of this, many astronomers visit Antarctica to collect data and observe the night sky. Since there are no lights, pollution, or smog to cloud the view of the sky, scientists often find that they're able to see the stars more clearly in Antarctica than nearly anywhere else in the world.

Sometimes, to get their instruments as high in the Antarctic sky as possible, scientists inflate large balloons to carry the instruments. The balloons can reach heights of 115,000 feet (35,000 m). The South Pole Telescope has also been set up in Antarctica to help scientists study space from this remote part of the world.

These instruments help astronomers study the most extreme things about space, such as the Big Bang that's believed to have formed our universe.

SCIENCE IN ACTION

The U.S. Antarctic ballooning program was formed through a partnership between the NSF and the National Aeronautics and Space Administration (NASA).

Because Antarctica is the location of the South Pole, there are some times during the year when it has 24 hours of daylight. At other times, there are 24 hours of darkness. This can be a challenge for many kinds of Antarctic researchers.

ANTARCTICA ROCKS!

Studying geology, or the science of Earth's history as seen through its land, can be difficult in Antarctica. Researchers must dig below the sheet of ice to reach the rocks they want to study! Oftentimes, they must drill in order to collect core samples.

Geologists study the mountains that rise above the Antarctic ice, such as the Transantarctic Mountains. Looking at the types of rocks and minerals found in the mountains can help scientists make **theories** about the types of life that existed on Antarctica in the past. For example, geologists working for the British Antarctic Survey (BAS) are using computer models to show what Antarctica may have looked like 3 million years ago. These researchers used ships and aircraft to reach the most remote corners of the continent to collect data.

SCIENCE IN ACTION

Many fossils can be found in the layers of rock below the surface of Antarctica. These fossils have helped scientists discover that dinosaurs once roamed the flat plains of Antarctica. Fossils show that the climate of Antarctica was much warmer millions of years ago.

Antarctica is one of the most extreme places on Earth to look for fossils!

RESEARCH STATIONS

Antarctic researchers have a unique job because—unlike many scientists—they don't work in a traditional lab. However, Antarctic geologists, biologists, astronomers, and other scientists need somewhere to live and to study their data and samples. Research stations, or bases, have been set up across the continent to allow scientists to have a safe and warm place to go when they aren't out in the field.

The BAS's Halley Research Station houses about 16 people during the winter and about 70 people during the warmer summer months. It's made up of eight separate, movable units.

SCIENCE IN ACTION

The Halley Research Station is sometimes known as Halley VI because it's the sixth version to be built by the BAS. The first five were either destroyed by snow or unable to be moved when conditions became dangerous. This is why Halley VI has movable units.

Engineers must be especially mindful of the materials they use to construct Antarctic research stations, because the cold and ice can damage the buildings.

Halley VI

AT HOME ON THE BASE

Antarctic research stations often have separate sections for working and living. Scientists can stay there for months at a time, so they make themselves at home. Since the purpose of these bases is for scientific research, most of the scientists' time is spent doing just that. Going to a movie theater or to the grocery store is impossible, because those kinds of buildings don't exist in such a cold place. Scientists must be careful to **ration** their supplies so they don't run out.

Since the bases are in such remote areas, it's also important that each has a doctor. This way, if someone gets sick or hurt, they can get proper medical care. Most bases also have an engineer to fix any mechanical or building issues.

SCIENCE IN ACTION

Some bases are in such remote areas that the only way to get to them is by helicopter.

Most Antarctic research bases are close to the shore, because that's where the weather is warmer and where most plants and animals live.

Antarctica

key

• research station

A DAY IN THE LIFE

Antarctic researchers have an exciting job! A typical day could include bundling up to observe wildlife or taking core samples to study rocks under the ice. The warmer months are generally busier, since that's when animals are more active. Depending on the type of research, some scientists may even set up a temporary camp so they can follow the species they're studying or be closer to a specific area of land.

Researchers also leave some time for recreation. Fun activities include photographing the beautiful landscape and animals, snowshoeing, skiing, snowmobiling, and ice fishing. Many scientists take advantage of these activities, since—depending on what part of the world they're from—they may not be able to do them at home.

SCIENCE IN ACTION

Scientists aren't the only people who travel to Antarctica. People can take tours of the continent. Over 30,000 of these tourists visit Antarctica every year.

Making time for fun in
Antarctica gives scientists a
break from their research.

BECOMING AN ANTARCTIC RESEARCHER

If all this sounds interesting and fun to you, perhaps you'd like to become an Antarctic researcher. There are a few things you should start doing now to prepare for this unique career.

First, learn as much as you can about the type of research you're interested in. Perhaps you want to study animals such as penguins or whales. Maybe you're interested in weather patterns or Antarctic storms. Check out books from your local library or ask an adult to help you research online to learn about your area of interest.

In college, study a kind of science you enjoy learning about, such as biology, geology, or astronomy. In order to become an Antarctic researcher, you'll need to work hard in school, especially in science classes.

SCIENCE IN ACTION

Antarctic researchers also need to have a good understanding of math. The data they collect—such as population figures, temperatures, and amounts of snowfall—is often shown through numbers.

Antarctic researchers should have knowledge of the latest technology used in whatever branch of science they're dealing with.

EXTREME SCIENCE

Being an Antarctic researcher certainly isn't boring! Although the conditions can be harsh and dangerous, many researchers say the experience of observing and collecting data on this unique continent is very interesting, rewarding, and fun. Without the hard work of these scientists, we wouldn't know nearly as much as we do now about Antarctica, its history, and how it influences the rest of the world.

If being an Antarctic researcher sounds fun to you, start researching! Maybe one day, you'll be living on a base and making important scientific discoveries in one of the most extreme places on Earth.

GLOSSARY

accumulate: To gather or build up.

blubber: The fat on whales and other large marine mammals.

complicated: Hard to understand.

deter: To cause someone to decide not to do something.

krill: Very small ocean creatures that are the main food for some whales.

logistics: The handling of the details of an operation.

marine biologist: A scientist who studies the living things in and around oceans.

microchip: A group of tiny electronic circuits that work together on a very small piece of hard material.

migration: The movement of animals from one place to another as the seasons change.

ration: To control the amount of something people are allowed to have, especially when the supply is limited.

sea level: The average height of the sea's surface.

theory: A working principle for which there is experimental evidence, but which has not been absolutely proven.

treacherous: Not safe.

unique: Special or different from anything else.

INDEX

A

Amundsen-Scott South Pole Station, 5

astronomers, 18, 22, 29

B

BAS, 20, 22

biologists, 14, 22, 29

C

climate, 6, 8, 10, 12, 13, 15, 20

core samples, 12, 20, 26

D

desert, 8

F

fossils, 20, 21

G

geology, 10, 20, 22, 29

H

Halley Research Station, 22, 23

I

ice, 8, 9, 12, 20, 22, 26

M

marine biologists, 16

math, 29

McMurdo Station, 10, 11

N

NASA, 18

NSF, 14, 18

P

penguins, 14, 15, 28

protective gear, 6, 7

S

sea levels, 12

snow, 8, 10, 22, 28

South Pole, 5, 14, 18, 19

T

telescope, 18

tourists, 26

Transantarctic Mountains, 20

U

U.S. Antarctic Program, 10, 14

W

Weddell seal, 14, 15

whales, 14, 16, 17, 28

wildlife, 10, 14, 26

WEBSITES

Due to the changing nature of Internet links, PowerKids Press has developed an online list of websites related to the subject of this book. This site is updated regularly. Please use this link to access the list:
www.powerkidslinks.com/exsci/anres